Text, photography, and illustrations © 2016 by The Taunton Press, Inc.

First published in this format 2016

Text: Rayan Turner
Cover and Interior Design: Kimberly Adis
Photographers: Alexandra Grablewski, except for photos on pp. 2, 4–5, 9, 11, 15, 21, 33, 35, and 39 by Rayan Turner
Illustrator: Rayan Turner
Editor: Tim Stobierski
Copy Editor: Valerie Cimino

The Taunton Press
Inspiration for hands-on living®

The Taunton Press, Inc., 63 South Main Street, PO Box 5506, Newtown, CT 06470-5506
e-mail: tp@taunton.com

Threads® is a trademark of The Taunton Press, Inc., registered in the U.S. Patent and Trademark Office.

The following names/manufacturers appearing in *3D Pen Jewelry* are trademarks: 3Doodler®

Library of Congress Cataloging-in-Publication Data

Names: Turner, Rayan.
Title: 3D pen jewelry : 20+ modern projects to make / Rayan Turner.
Description: Newtown, CT : The Taunton Press, Inc., 2016. | Includes bibliographical references and index.
Identifiers: LCCN 2016030941 | ISBN 9781631867101 (alk. paper)
Subjects: LCSH: Plastics craft. | Plastic jewelry. | Three-dimensional printing. | Handicraft for girls.
Classification: LCC TT297 .T87 2016 | DDC 745.57/2--dc23
LC record available at https://lccn.loc.gov/2016030941

Printed in the United States of America
10 9 8 7 6 5 4 3 2 1

Contents

Introduction

3D PENS ARE TAKING THE CRAFT WORLD by storm. The perfect tool to add to your arsenal, a 3D pen will take your projects to the next level. Though it takes some practice to learn to use the pen, once you are accustomed to how it works, the possibilities of what you can create are endless. The beauty of using a 3D pen to make jewelry is that you can create dimensional designs that mimic metalwork or machined pieces, previously possible only for skilled craftsmen or those involved in high-tech manufacturing. An off-the-charts heartbeat (p. 40), out-of-this-world constellations (p. 16), and beautiful, natural, organic designs are just a few possibilities. If you are new to using 3D pens, thoroughly read the manufacturer's instructions and play around until you understand how the tool works. There are a number of brands available, and each works slightly differently; for most of these projects, I used the 3Doodler®. The following pages will give you an introduction to the tools, materials, and techniques used to create the jewelry projects in this booklet.

Tools and Materials

TO MAKE THE PROJECTS IN THESE PAGES, you will need a 3D pen, filament, and basic jewelry-making tools and materials. Some brands of 3D pen come with accessories like different nozzles or mats specifically designed for use with a 3D pen, which may come in handy.

FILAMENT

Filament comes in a huge range of colors and materials, and depending on the brand of pen you use, it may come in strand or spool form. The 3 basic types of filament are ABS, PLA, and flexible filament. ABS is a petroleum-based plastic that is hard, impact resistant, and tough. It maintains some flexibility, making it ideal for jewelry projects. PLA is a plant-based plastic that is similar to ABS, but it can stick directly to fabric, other plastics, metal, or glass; it is also rigid when cool and perfect for translucent designs. Flexible filament is essentially a rubber, and remains flexible and pliable even after it cools, making it the perfect choice for projects that will wrap around the wrist or neck.

BASIC JEWELRY TOOLS

The tools you will need for these projects are those you'd need for any jewelry project: scissors, for snipping away excess filament; a wire cutter to cut necklace chain; a pair of small pliers to manipulate hardware; paper, for tracing your templates onto; and jewelry cement or glue to secure your projects to hardware. A heat-resistant silicone thimble can be extremely helpful but is not required for any of these projects.

JEWELRY HARDWARE

All of these projects can be attached to any type of basic jewelry hardware. The hardware used in these projects includes: necklace chain, which comes in many styles and thicknesses; leather, suede, or cotton cord; jump rings (4mm, 6mm, and 8mm), which are used to connect the projects to the hardware; fold-over cord ends, which fold over the cut ends of cord, chain, or filament and allow you to attach additional hardware; crimp beads, which are used to secure the ends of wire or string once it has been attached to a clasp or jump ring; clasps, which come in a number of styles; bobby pins; and French wire earring hooks. You can also easily attach the projects to bracelet or ring blanks, key chains, or any other hardware that you can think of.

Working with Your 3D Pen

THOUGH EACH BRAND OF PEN WORKS differently, here are a few tips and tricks that I've learned while using my 3D pen that will help you quite a bit in these projects. 3D pens are a lot like cats and toddlers: They might not always work the way you want them to at first, but as long as you practice patience, everything will work out.

▶ Filament sticks to paper, so it is very important to always trace, copy, or clip a tracing mat to the templates in this booklet. If you don't, you will likely tear the page when peeling away the finished project.

▶ It's easy to change the size of any of the templates in this booklet—just use a photocopier.

▶ If your pen comes with different-sized nozzles, always use the largest nozzle for these projects. The thicker the extruded filament, the stronger the final piece will be.

▶ When beginning a project, give yourself a running start by beginning to extrude the filament before pressing the tip of the 3D pen to your template. This will remove any air bubbles or hardened filament from previous use and will make the project so much easier to complete.

▶ I've indicated in the Materials lists which projects require flexible filament—these are projects that need to stretch or bend, or that will receive a lot of wear and tear. In cases where this is not specified, you can use any filament of your choice.

▶ If your filament is not sticking to the paper or tracing mat, move the pen more slowly and apply more pressure as you trace.

▶ If you run into any issues with how your pen operates, always refer back to the manufacturer's instructions or contact the manufacturer directly; most are happy to guide you through a troubleshooting process.

Techniques

WORKING WITH A STANDARD NECKLACE CHAIN WITH CLASP

The steps below show an easy way to prep a necklace chain for attachment to your 3D pen projects.

1. Use wire cutters to cut the chain opposite the clasp/closure.

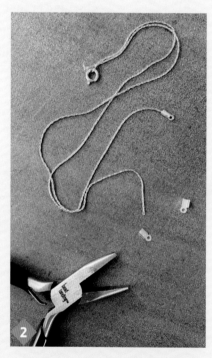

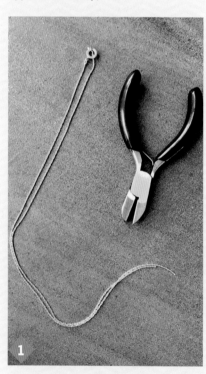

2. On each end of the cut chain, attach fold-over cord ends by crimping them in place with pliers. Once the fold-over cord ends are in place, use pliers to attach a jump ring to each cord end. Then, attach the jump rings to your design. See the Drip Drop Necklace on p. 32 for an example of a project that uses this technique.

WORKING WITH BULK CHAIN OR CORD

If you are working with bulk chain or cord, it will not have a clasp or closure, so you will need to create one yourself using the steps below.

1. Cut the chain to the desired length using wire cutters. Cut this length in half.

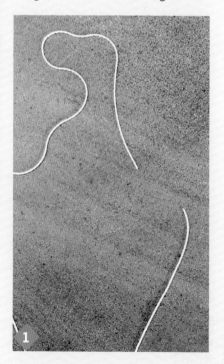

2. On each of the 4 ends of the cut chain, attach fold-over cord ends by crimping them in place with pliers. Once the fold-over cord ends are in place, use pliers to attach a 4mm jump ring to 3 of the cord ends.

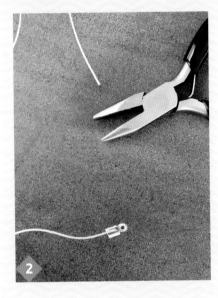

3. Finish the closure by attaching a lobster clasp (or closure of your choice) to one of the 4mm jump rings. On the remaining cord end, attach a 6mm or 8mm jump ring. The lobster clasp will partner with the larger jump ring. Use pliers to attach the 2 remaining 4mm jump rings to the design. See the Ruffled Feathers Necklace on p. 22 for an example of this technique.

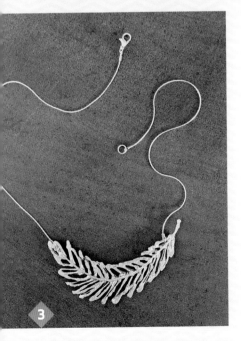

WORKING WITH FRENCH WIRE EARRING HOOKS

The earring projects in this booklet all use French wire earring hooks for hardware. For dangle earrings, simply attach an 8mm jump ring to the loop on the earring hook and then attach to your design (see the Infinitely Circular Earrings on p. 12 for an example).

For projects that require the design to sit vertically along the hook, such as the Listen to Your Heartstrings Ear Climber on p. 36, you will need to follow the steps below to attach the hardware to your design without using a jump ring.

1. Use a pair of pliers to bend the small loop at the end of the hook so that it sits 90 degrees perpendicular to the rest of the hook.

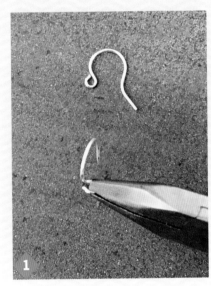

2. Use the photo below as reference for what the hook will look like after the loop has been bent.

3. After the loop is bent, simply cinch the loop closed snugly around the jewelry design. When the project sits vertically along the length of the hook, it is best if the loop starts ¼ in. from the edge of the design; this allows a portion of the jewelry to hide the top of the hook from view, as shown in the photo below.

Crescent Moon Pendant

Shoot for the moon—even if you miss, you will land among the stars. This pendant is the perfect simple accessory, looking great in white or in a metallic finish. To really play up the fun side of the moon design, you can finish with a metallic paint or use a glow-in-the-dark filament that will let you shine!

MATERIALS

3D pen

Filament in a color of your choice

Scissors

Metallic paint in a color of your choice (optional)

4mm jump ring

Necklace chain with clasp

Pliers

TO MAKE THE PENDANT

1. Copy the template on p. 44 onto plain paper. Using the largest nozzle on your 3D pen, begin extruding the filament—this will give you a running start—and then press the tip of the 3D pen to your template. Trace the template using a slow and steady pace. Keep the nozzle at a slight angle and your hand elevated above the work surface.

2. When you arrive back at the starting point, continue extruding the filament past the starting point of the moon for an additional ¼ in. This will help you achieve a nice joint with ends that will be easy to trim away and perfect later. Inspect your work and touch up any areas as necessary. Once the filament is completely cooled, gently peel the design from the template and trim away the excess filament with the scissors. If you would like to paint the pendant in a metallic finish, now is the time to do so; I used a silver paint.

3. To make the pendant as shown, use the pliers to open the jump ring. Close the jump ring around one tip of the crescent moon, and insert the length of chain into the jump ring.

variation

Instead of making a pendant, make a bracelet by attaching a jump ring to both sides of the crescent moon and linking the jump rings with a length of chain or cord that comfortably fits around your wrist.

The Cat's Meow Ring

These adorable rings take their inspiration from our favorite four-legged friends, making them the perfect gift or accessory for the cat lover in your life. Go ahead—make one for each of your nine lives.

MATERIALS

3D pen

Flexible filament in a color of your choice

Scissors

TO MAKE THE RING

1. Measure your finger by wrapping a piece of string around it and then placing the string on the chart on p. 46 to determine your ring size. Copy the correct template on p. 44 onto a sheet of paper.

Note: It is best to begin tracing the template at the bottom of the ring so that once you have finished tracing the template, the connecting point in the ring is hidden when the ring is worn.

2. Using the largest nozzle on your 3D pen, begin extruding the filament—this will give you a running start—and then press the tip of the 3D pen to your template. Trace the template using a slow and steady pace. Keep the nozzle at a slight angle and your hand elevated above the work surface.

3. When you arrive back at the starting point, continue extruding the filament past the starting point of the ring for an additional ¼ in. to create a solid joint. Inspect your work and touch up any areas as necessary. Once the filament is completely cooled, gently peel the design from the template and trim away the excess filament with the scissors.

> ### variation
> Use these templates to create other types of accessories besides rings. Enlarge the template slightly using a photocopier to create a custom tie for your ponytail, or go bigger for a bracelet that fits comfortably around your wrist.

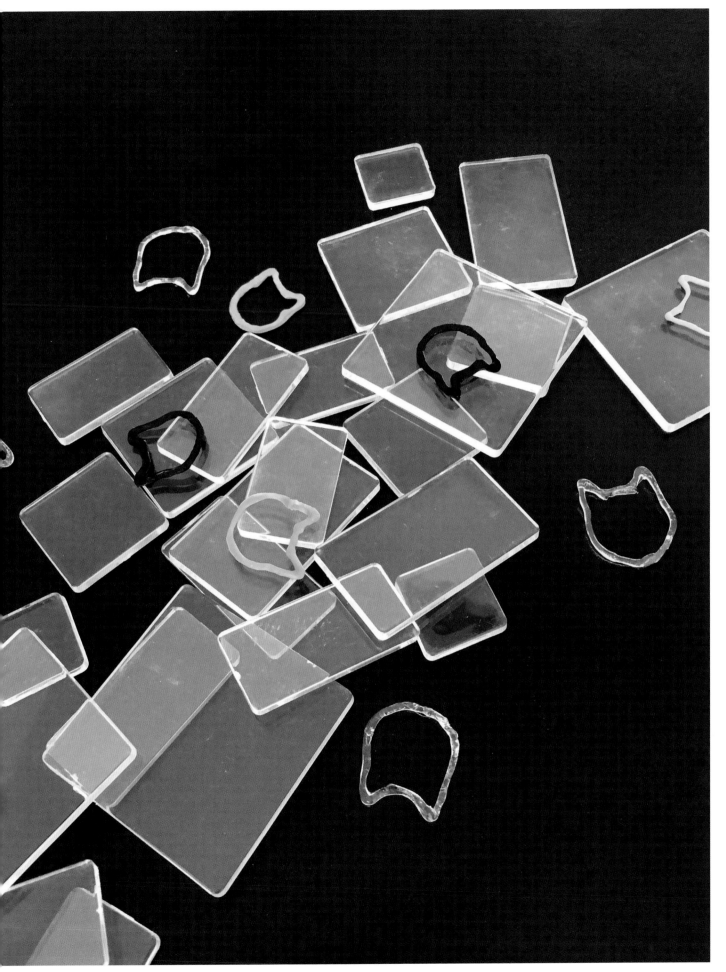

Modern Coral Lariat

Someone once said that "our memories of the ocean will linger on long after our footprints in the sand are gone." This sleek lariat was inspired by the sea—a modern take on a coastal classic.

MATERIALS

3D pen

Flexible filament in a color of your choice

Scissors

Pliers

Wire cutters

Two 4mm jump rings

4 fold-over cord ends

21 in. of bulk chain or cord

TO MAKE THE LARIAT

1. Copy the coral templates on p. 44 onto plain paper. Using the largest nozzle on your 3D pen, begin extruding the filament—this will give you a running start—and then press the tip of the 3D pen to your template. Trace the center stem of the larger template using a slow and steady pace. Keep the nozzle at a slight angle and your hand elevated above the work surface.

Note: Overlap each leaf with the stem multiple times for a stronger bond.

2. Once the stem is complete, move onto each leaf, working them one at a time; each leaf should overlap the center stem. Once each leaf is outlined, fill it in with filament in a circular motion. Make sure each leaf is properly secured to the stem. Inspect your work and touch up any areas as necessary. Once the filament is completely cooled, gently peel the design from the template and trim away the excess filament with the scissors.

3. Repeat steps 1 and 2 with the smaller template so that you have 2 pieces of coral for the lariat.

4. To make the lariat as shown, use pliers to crimp a fold-over cord end to each coral stem, and also to each end of your chain or cord. (You may need to trim the stem slightly after it is inserted into the cord end.) Use the pliers to open each jump ring. Insert a jump ring into the fold-over cord end on each end of the chain, and then link each jump ring to one of the stems. See p. 4 for more detail on working with bulk chain. Wear the lariat by placing the necklace on your neck and looping the ends as if tying the first half of a square knot (or your shoelaces).

variation

To make cute coordinating earrings, reduce the template to a comfortable size and simply secure an earring hook to the stems as directed on p. 5.

Infinitely Circular Earrings

MATERIALS

3D pen

Filament in a color of your choice

Scissors

2 French wire earring hooks

Two 8mm jump rings

Pliers

Hoop earrings are a fashion staple, but why settle for one hoop when you can have three? Use a metallic gold finish for the classic look, swirl away in your favorite color for a statement, or alternate colors for an amazing ombré effect.

TO MAKE THE EARRINGS

1. Copy the template on p. 44 onto plain paper. Using the largest nozzle on your 3D pen, begin extruding the filament—this will give you a running start—and then press the tip of the 3D pen to your template. Trace the innermost (smallest) circle of the template using a slow and steady pace. Keep the nozzle at a slight angle and your hand elevated above the work surface. Work outward until you have completed all 3 circles.

2. To create the dimensional effect that these earrings show, trace the 2 inner circles twice and the outer (largest) circle 3 times. At the top of the earring, where the circles overlap, the largest circle will sit on top of the middle circle, which sits on top of the smallest circle.

3. Repeat steps 1 and 2 to create a second earring. Inspect your work and touch up any areas as necessary. Once the filament is completely cooled, gently peel the design from the template and trim away the excess filament with the scissors.

4. To add hardware to the design, use pliers to open the jump rings and attach one jump ring to the bottom of each earring hook. If the jump ring is large enough, simply enclose it around the top of the earring. If it is not large enough, and you used a flexible filament, you should be able to maneuver the jump ring through the layers of filament. See p. 5 for more detail on working with earring hooks.

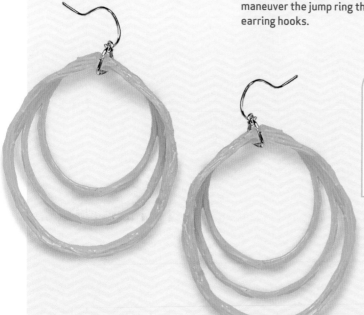

variation

To make a pendant, simply trace each circle once and attach a jump ring as in the earrings above. Instead of attaching the jump ring to an earring hook, string the piece onto a necklace chain or cord.

Arrowhead Pendant

MATERIALS

3D pen

Flexible filament in a color of your choice

Scissors

Pliers

Necklace wire or chain with clasp

4mm jump ring

You don't have to make a trip to the desert to score an arrowhead to fashion into a pendant: Make one for yourself! Wear this timeless design and channel your inner excavator—join the ranks of those who wander and seek adventure.

TO MAKE THE PENDANT

1. Copy the template on p. 45 onto plain paper. Using the largest nozzle on your 3D pen, begin extruding the filament—this will give you a running start—and then press the tip of the 3D pen to your template. Trace the template beginning with the top (straight portion) of the template. When you are done with the top line, without lifting the pen, move down and make a second line right next to the first. Continue working back and forth in this manner until you have filled in the template. Each line should rest snugly against the previous line to prevent gaps.

2. Inspect your work and touch up any areas as necessary. Once the filament is completely cooled, gently peel the design from the template and trim away the excess filament with the scissors. Turn the arrowhead over to the back side and fill in any gaps as needed.

3. To make the pendant as shown, use pliers to open a 4mm jump ring and thread it through the layers at the top of the arrowhead. If it is too difficult to do so, you can use a safety pin to poke a hole through the lines of filament. Then, close the jump ring and thread it onto the necklace wire or chain.

> ## variation
> To make a bobby pin or hair clip instead, simply glue the arrowhead onto a bobby pin (see the Banter Bobbies on p. 18 to see instructions for a similar project). Alternatively, make a sleek bracelet by using a length of chain that comfortably fits around your wrist.

What's Your Sign Pendant

MATERIALS

3D pen

Filament in a color of your choice

Scissors

Pliers

Wire cutters

26-gauge jewelry wire

Two 4mm jump rings

2 crimp beads

Clasp

8mm jump ring

What does your constellation say about you? This pendant can let you tell the world who you are, or carry the star signs of your favorite people close to your heart. Star gazing takes on a whole new meaning when you can look to the heavens for some celestial inspiration.

TO MAKE THE PENDANT

1. Copy a template from p. 45 onto plain paper. Select a "star" at one end of the template; this will be your starting point. Using the largest nozzle on your 3D pen, press the tip of the pen firmly against the template at your starting point and hold it in place for several seconds while the filament extrudes and forms a rounded ball. This will give dimension to the star.

2. When the star has the desired dimension, trace the connecting line of the template to the next star and repeat the process. Continue around the template until the design is complete.

Note: If the constellation you choose is a closed design, begin by filling in the first star only slightly. You will finish adding dimension to the star when you arrive back at this point at the end of the process.

3. Inspect your work and touch up any areas as necessary. Once the filament is completely cooled, gently peel the design from the template and trim away the excess filament with the scissors.

4. To make the pendant as shown, cut two 20-in. lengths of jewelry wire with wire cutters. Fold each length in half and string a 4mm jump ring onto each, allowing it to slide down into the crook at the bend of each wire. Working with one of the lengths of bent wire, hold the 2 ends of the wire together and slide a crimp bead onto the end. Then, thread the 2 ends of wire through the loop of the clasp and fold them back into the crimp bead. Crimp the bead closed using pliers, and repeat for the second wire, working with an 8mm jump ring instead of a necklace clasp.

5. Use pliers to open the 4mm jump ring at the bent end of one of the lengths of wire. Place it around one end of the constellation and close securely. Repeat on the other side with the second jump ring.

variation

You can use these templates to create necklaces, bracelets, or dangle earrings—many different kinds of jewelry. Experiment with hardware to see what works for you.

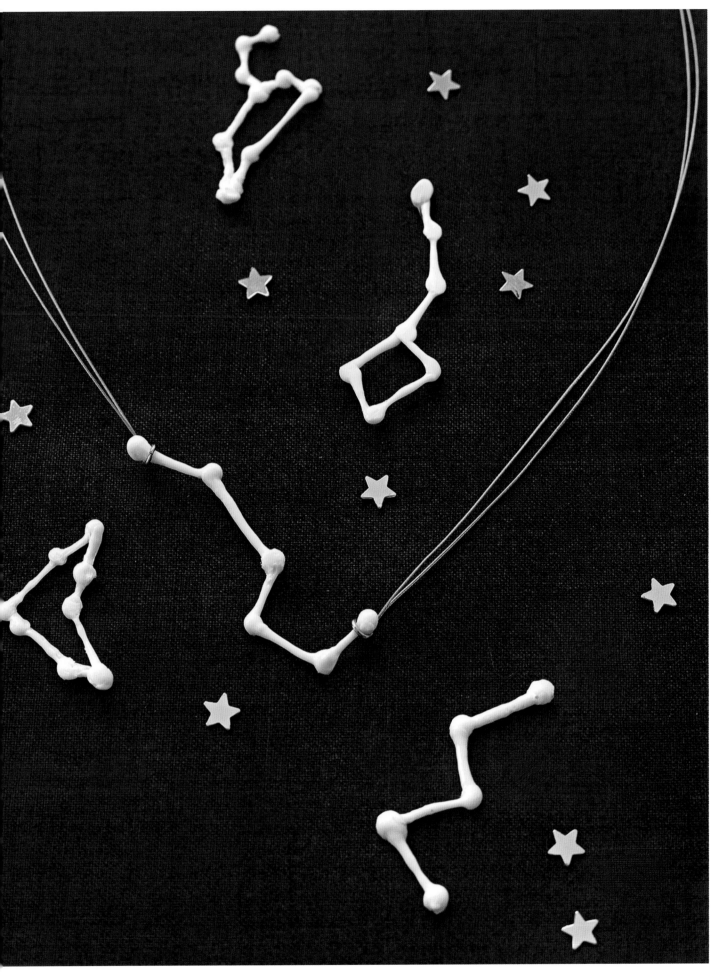

Banter Bobbies

Now you can speak your mind without uttering a single word. Use these cute and sassy designs to scream your individuality no matter the time or place.

MATERIALS

3D pen

Filament in a color of your choice

Scissors

Bobby pins

Pliers

Jewelry cement or glue

TO MAKE THE BOBBY PINS

1. Copy the template on p. 45 onto plain paper. Using the largest nozzle on your 3D pen, begin extruding the filament—this will give you a running start—and then press the tip of the 3D pen to your template. Trace the template using a slow and steady pace. Keep the nozzle at a slight angle and your hand elevated above the work surface.

2. Trace each letter separately, continuing a bit past the end point (you will trim this later); simply overlap the ends if you've chosen a closed letter. When you have traced each letter once, go over it a second time to add dimension and strength to the design. Inspect your work and touch up any areas as necessary. Once the filament is completely cooled, gently peel the design from the template and trim away the excess filament with the scissors.

3. To make the bobby pins as shown, secure each bobby pin to your work surface (you can hold the bobby pin in place by clamping it to your work surface with a binder clip). Be sure the top edge or surface is facing upward. Run a thin line of jewelry cement or glue along the portion of the bobby pin where the letters will sit. Place the letters on the bobby pin using pliers and adjust them so that each letter sits centered from top to bottom and spaced consistently. Allow the glue to set according to the manufacturer's instructions.

variation

Instead of making a bobby pin, you can easily attach the letters to a bracelet blank or key chain. Or, instead of forming the letters individually, simply have them touch on the sides to form one solid piece.

When in Rome Bangle

MATERIALS

3D pen

Flexible filament in a
color of your choice

Scissors

Metallic paint in a
color of your choice
(optional)

4 fold-over cord ends

Four 4mm jump rings

Two 6mm or 8mm
jump rings

2 clasps

Pliers

When in Rome, do as the Romans do. This modern design is minimal and sleek and requires fairly little hardware to complete. Trace the numerals in order, or customize with a number that has personal meaning to always keep it with you.

TO MAKE THE BRACELET

1. Copy the template on p. 46 onto plain paper. Measure your wrist, and mark the template at that length; if you need to make the template longer, just repeat the design until you reach your desired length.

2. Using the largest nozzle on your 3D pen, begin extruding the filament—this will give you a running start—and then press the tip of the 3D pen to your template. Trace the first Roman numeral of the bangle, continuing past the ends of the numeral by ¼ in. (you will trim these later). Repeat for all of the numerals in the template. Inspect each numeral and reinforce as necessary. Without removing the numerals from the template, snip away any excess filament with the scissors, making sure each numeral is the same height.

3. Trace the top band of the bracelet, moving slowly and making sure to overlap the edge of each numeral. Repeat for the bottom edge of the bracelet. Inspect your work and touch up any areas as necessary. Once the filament is completely cooled, gently peel the design from the template and trim away the excess filament with the scissors. If you wish to paint the bracelet in a metallic finish, you may do so now; I used a metallic bronze paint.

4. To make the bracelet as shown, slide a fold-over cord end onto each end of the bracelet. Be sure the flat side of the fold-over cord end is facing upward. On one side of the bracelet, use pliers to attach a 4mm jump ring to each fold-over cord end. On the opposite side, attach a 6mm or 8mm jump ring to each cord end, and then attach a clasp to each of those. To wear, latch the clasps to the corresponding jump rings.

> ### variation
> You can easily make a headband or choker-style necklace by increasing the template to a length that fits comfortably.

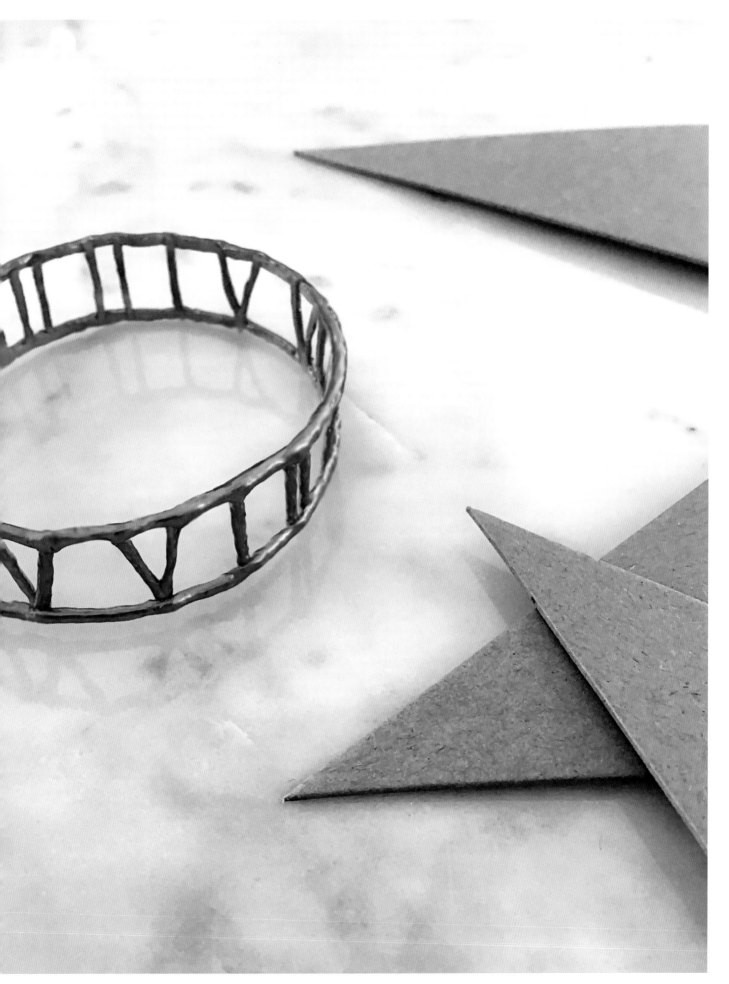

Ruffled Feathers Necklace

MATERIALS

3D pen

Filament in a color of your choice

Scissors

Metallic paint in a color of your choice (optional)

Pliers

Wire cutters

Necklace chain with clasp

Two 4mm jump rings

2 fold-over cord ends

Birds of a feather flock together, but aren't afraid to stand out from the crowd. Go ahead—ruffle some feathers with this stylish statement piece. Paint it in a metallic finish like I did, or rock your favorite colors.

TO MAKE THE NECKLACE

1. Copy the template on p. 46 onto plain paper. Using the largest nozzle on your 3D pen, begin extruding the filament—this will give you a running start—and then press the tip of the 3D pen to your template. Trace the centerline of the feather template using a slow and steady pace. Keep the nozzle at a slight angle and your hand elevated above the work surface. Continue about 1 in. past the end of the template (you will trim this to the desired length after you are done attaching hardware).

2. To create the feather portion of the design, work one half of the feather in its entirety before moving on to the second half. Begin extruding filament close to the centerline, but not directly onto it. Then move the pen toward the centerline so that you overlap it, and then work back away from the centerline. Overlap the centerline like this for each individual tuft of the feather. Repeat for the other side of the feather.

3. Inspect your work and touch up any areas as necessary. Once the filament is completely cooled, gently peel the design from the template and trim away the excess filament with the scissors. Try not to stress any connecting points when removing the design. If you would like to paint the pendant in a metallic finish, now is the time to do so; I used a silver paint.

4. To make the necklace as shown, cut the necklace chain in half opposite the clasp using the wire cutters. Attach a fold-over cord end to each end of the length of chain. Then use pliers to attach a 4mm jump ring to each fold-over cord end. On each end of the feather design, locate an open spot in the design that can accommodate a jump ring, and attach the jump rings at these spots. See p. 4 for more detail on working with necklace chain.

variation
Use the smallest feather templates on p. 46 to create a matching set of earrings, or reduce the larger template a bit to create a stylish bracelet or headband.

All Coiled Up Ring

MATERIALS

3D pen

Flexible filament in a
color of your choice

Scissors

Metallic paint in a
color of your choice
(optional)

Wrapped wire and coil rings are all the rage: Elegant and sophisticated, yet organic and a bit unruly, they are perfect for the girl who is equally at ease in the great outdoors or all dressed up for a night on the town.

TO MAKE THE RING

1. Measure your finger and use the size chart on p. 46 to determine your ring size (as directed in step 1 on p. 8). Copy this ring size onto plain paper; it will be your template for the project. Using the largest nozzle on your 3D pen, begin extruding the filament—this will give you a running start—and then press the tip of the 3D pen to your template. Trace the template using a slow and steady pace. Keep the nozzle at a slight angle and your hand elevated above the work surface.

Note: If your finger doesn't fit any of the ring sizes perfectly, choose the template that best fits and adjust by tracing along the inside of the template for a slightly tighter fit or along the outside for a looser fit.

2. When you return to the starting point of the template, simply lift the pen slightly and continue onto the first layer of filament. Work the second layer in the same thickness as the first. Repeat for as many layers as desired. Do not be discouraged if the layers are not identical—that very quality is what gives the design its coiled-wire look.

Note: If you notice that a segment of the ring is lower than the others (sometimes the filament settles in odd ways), then simply slow down your pace when you reach the lower section so that the section builds up to match the rest of the ring.

3. Inspect your work and touch up any areas as necessary. Once the filament is completely cooled, gently peel the design from the template and trim away the excess filament with the scissors. If you would like to paint the pendant in a metallic finish, now is the time to do so; I used a bronze paint. Wear the ring so that its base (which will be flat) sits at the base of your finger.

variation

Enlarge one of these templates substantially to create a unique hair accessory that you can use to cover your ponytail holder.

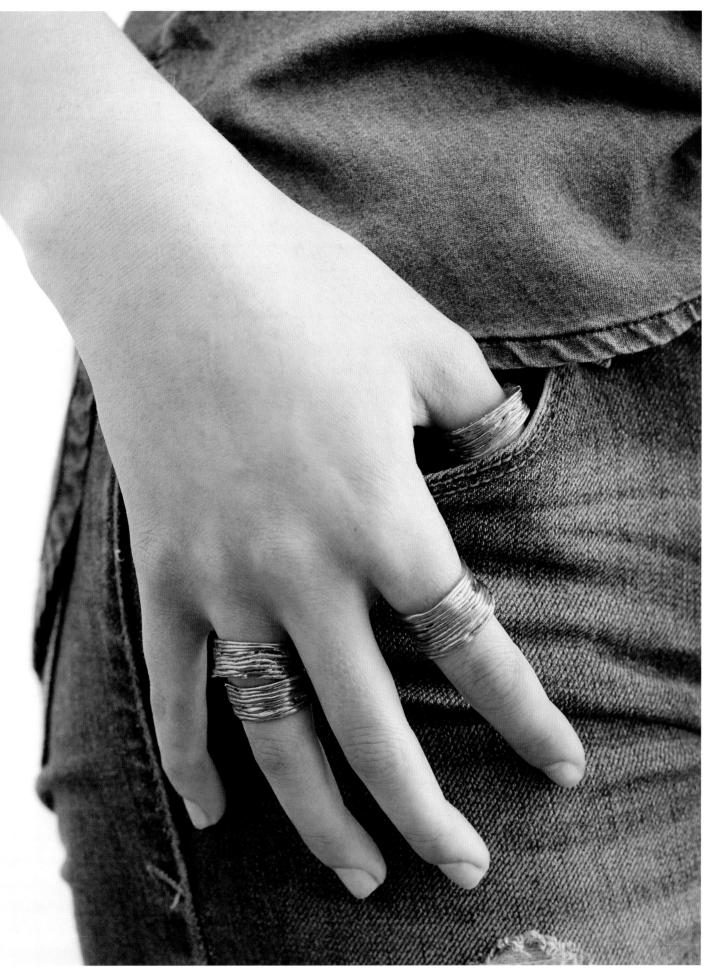

Geo-Totem Necklace

This pendant takes geometry to the extreme. But don't worry, there isn't any tricky math involved in the project—just a few simple overlapping rectangles and circles. Paint the piece with a metallic finish for an elegant hammered metal look, or rock out in bright and bold neon.

MATERIALS

3D pen

Filament in a color of your choice

Scissors

Jewelry cement or glue

Paint in a color of your choice (optional)

Pliers

Wire cutters

Two 4mm jump rings

2 fold-over cord ends

Necklace chain with clasp

TO MAKE THE NECKLACE

1. Copy the 2 templates on p. 47 onto plain paper. Using the largest nozzle on your 3D pen, begin extruding the filament—this will give you a running start—and then press the tip of the 3D pen to your template. Trace the template using a slow and steady pace. Keep the nozzle at a slight angle and your hand elevated above the work surface.

2. Trace the bottom 3 sides of the 2 larger rectangles first, continuing slightly beyond the template. Follow the process for the smaller rectangles, using the bottom edge of the larger rectangle as the top edge of the 2 small rectangles. Then, trace the circle templates, beginning with the inner circle. When the circles are complete, trace the straight line through the upper section of each. Touch up any portion of the design as needed. Without removing anything from the template, trim away excess filament with the scissors; for the circles, trim away the arc above the straight line.

3. Create the top edge of the larger rectangles by extruding filament over the top of the already created rectangle sides. Extend $\frac{1}{2}$ in. beyond both ends of the rectangle; this will be used to connect to the circular section.

4. Gently peel the rectangular and circular sections of the design away from the template. Inspect your work and touch up any areas as necessary. Position the rectangular section on top of the circular section, aligning the straight lines, and attach using jewelry cement or glue. Allow it to set according to the manufacturer's instructions. If you wish to paint the pendant, you may do so now; I used a metallic bronze paint.

5. To make the necklace as shown, cut the necklace chain in half opposite the clasp using wire cutters. Attach a fold-over cord end to each cut end of the chain. Then use pliers to attach a jump ring to each fold-over cord end. Loop each jump ring around opposite ends of the circular section of the design. See p. 4 for more detail on working with necklace chain.

It's Molecular Pendant

Some of the prettiest designs take their cue from nature, sometimes from things so small you can't see them with your naked eye. Who says that chemistry has to be boring?

MATERIALS

3D pen

Filament in a color of your choice

Scissors

Pliers

Wire cutters

Two 4mm jump rings

2 fold-over cord ends

Necklace chain with clasp

Jewelry cement

TO MAKE THE PENDANT

1. Copy a template on p. 47 onto plain paper. Using the largest nozzle on your 3D pen, begin extruding the filament—this will give you a running start—and then press the tip of the 3D pen to your template. Trace the top line of the template using a slow and steady pace to form the foundation of the pendant. Keep the nozzle at a slight angle and your hand elevated above the work surface. Continue a bit beyond the end of the template (you will trim this later). Trace the top line with another 2 layers of filament.

2. To create the hexagon shapes, start at the point where the hexagon touches the top line. Extrude so that the filament begins at the top line and then trace down the first side of the hexagon. Pause at the angle so that you have crisp corners, and then move around the rest of the shape. Be sure that the filament catches the top line when you return to it. Repeat this process for the rest of the design. Thicken the design by laying down another layer of filament.

3. Inspect your work and touch up any areas as necessary; take special care that the joints are strong. Once the filament is completely cooled, gently peel the design from the template and trim away the excess filament with the scissors.

4. To make the pendant as shown, use pliers to open both of the 4mm jump rings and thread each one through the eye of a fold-over cord end. Loop one jump ring around one end of the design, and close. Cut the necklace chain in half opposite the clasp using wire cutters. Insert one end of the necklace chain into the fold-over cord end and crimp it closed. Repeat on the other end. See p. 4 for more detail on working with necklace chain.

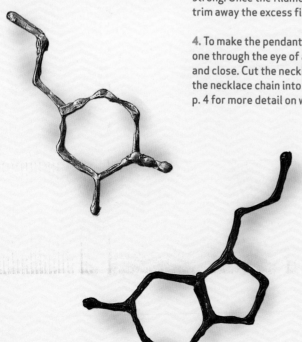

> ## variation
> Reduce your template by 50 percent using a photocopier. Attach it to a length of chain that comfortably fits around your wrist for a chic bracelet, or create 2 molecules—that mirror each other—for some fabulous earrings.

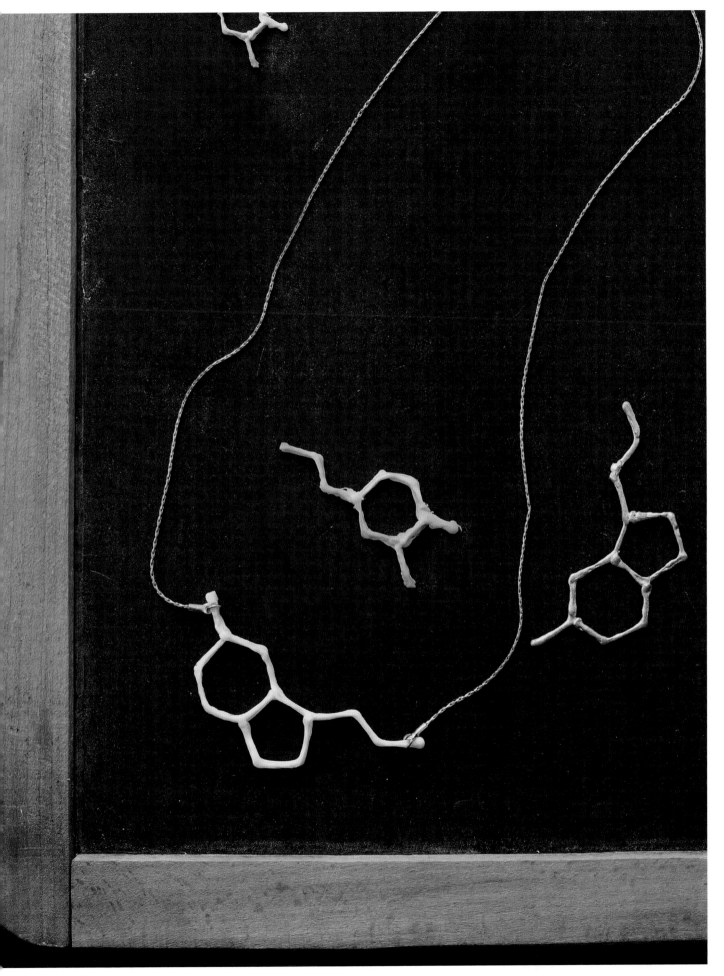

On the Vine Bracelet

This bracelet will take some time to make, but it's easier than it looks. Even if you don't have a green thumb, here's one plant that will stay green year-round.

MATERIALS

3D pen

Flexible filament in green

Scissors

2 fold-over cord ends

4mm jump ring

8mm jump ring

Clasp

Pliers

TO MAKE THE BRACELET

1. Copy the bracelet template on p. 47 onto plain paper. Measure your wrist, and mark the template to this length; if you need to make the template longer, just repeat the design until you reach the desired length. Using the largest nozzle on your 3D pen, begin extruding the filament—this will give you a running start—and then press the tip of the 3D pen to your template. Trace the stem of the template using a slow and steady pace. Keep the nozzle at a slight angle and your hand elevated above the work surface, and continue a bit beyond the end of the template (you will trim this later).

2. When making the leaves, it is easiest to create all of the leaves along one side of the stem before moving to the other side. Position the template so that you can work on the side of the stem opposite you. Begin extruding the filament prior to tracing the template. Catch the stem with your filament, and then trace the outline of the leaf; when you return to the base of the leaf, catch the stem in the same spot and stop extruding a bit beyond the stem (you will trim this later). Repeat this process, working inward in concentric rounds to fill in the leaf; each line of filament should rest snugly against the previous line to prevent gaps.

3. Repeat step 2 for all leaves on the side of the stem you are working on. Once the filament is cooled, but without removing the design from the template, trim away the filament beyond the stem. Repeat these steps for the other side of the stem.

4. Inspect your work and touch up as necessary. Once the filament is completely cooled, gently peel the design from the template and trim away the excess filament with the scissors. Be sure to loosen each leaf before trying to peel the vine off of the template, or you may break the stem. If any leaves do separate from the stem, simply use additional filament to reattach.

5. To make the bracelet as shown, attach a fold-over cord end to the stem at each end of the bracelet and crimp in place. Then, use pliers to open a 4mm jump ring and thread it through the eye of a fold-over cord end on one side of the bracelet. Open an 8mm jump ring and thread it through the eye of the other fold-over cord end. Attach the bar portion of the clasp to the 4mm jump ring, and attach the ring portion to the 8mm jump ring.

variation

To create a necklace, copy the necklace template on p. 47 to a length that is comfortable for your neck. Apply filament as described above, then attach hardware in the same manner as described above. Double the filament for the stem.

Drip Drop Necklace

This necklace's organic curves take their inspiration from flowing water. Use a translucent filament and become one with the water, or try using a colorful filament (or multiple colors) to really make a splash.

MATERIALS

3D pen

Flexible filament in a color of your choice

Scissors

Pliers

Wire cutters

Two 8mm jump rings

2 fold-over cord ends

Necklace chain with clasp

TO MAKE THE NECKLACE

1. Copy the template on p. 47 onto plain paper. Using the largest nozzle on your 3D pen, begin extruding the filament—this will give you a running start—and then press the tip of the 3D pen to your template. Trace the arc of the template using a slow and steady pace. Keep the nozzle at a slight angle and your hand elevated above the work surface. When you reach the end of the arc, trace it again to add stability and provide a solid foundation to the design.

2. Return to the starting point and trace the arc a third time. When you reach the first drip, follow the drip pattern by pulling the pen downward and making sure it is connected to the arc. Move slowly, so that the drip is thick. At the bottom of the drip, hold the pen in place and extrude for a few seconds to form the droplet. Continue extruding filament and return to the arc by pulling the pen upward; then make your way to the next drip. Repeat for all of the drips.

3. Inspect your work and touch up any areas as necessary. Once the filament is completely cooled, gently peel the design from the template and trim away the excess filament with the scissors.

4. To make the necklace as shown, use pliers to open an 8mm jump ring and loop it onto the arc just inside the first drip on each side. Cut the necklace chain in half opposite the clasp using wire cutters. On each end of the necklace chain, attach a fold-over cord end, and thread the jump ring through the eye of the fold-over cord end to complete. See p. 4 for more detail on working with necklace chain.

---

variation

Create a unique pendant by working this design with a single droplet, or attach the droplet design to the back of a hair clip for an interesting accessory.

Say It Out Loud Necklace

These funky necklaces tell it like it is and are perfect for a day of shopping, a day on the beach, and everywhere in between. Wear one, or layer them in multiples to spice up a casual outfit.

MATERIALS
3D pen

Flexible filament in a color of your choice

Scissors

Pliers

Two 4mm jump rings

4 fold-over cord ends

18 in. of necklace chain or cord

TO MAKE THE NECKLACE

1. Copy a template on p. 48 onto plain paper. Using the largest nozzle on your 3D pen, begin extruding the filament—this will give you a running start—and then press the tip of the 3D pen to your template. Trace the template using a slow and steady pace. Keep the nozzle at a slight angle and your hand elevated above the work surface. Trace the design at least 2 times to increase the durability and readability of the word; make sure that you work each layer of filament in the same direction for consistency.

2. Inspect your work and touch up any areas as necessary. Once the filament is completely cooled, gently peel the design from the template and trim away the excess filament with the scissors.

3. To make the necklace as shown, crimp a fold-over cord end to each end of your design. Then, secure a fold-over cord end to each end of your length of necklace chain or cord. Use pliers to open a 4mm jump ring and thread it through the eye of the fold-over cord and connect one end of the design with one end of the cord. Repeat for the other side. See p. 4 for more details on working with bulk cord.

variation
Use these templates to create cute bobby pins like the Banter Bobbies on p. 18, or try something different, like a key chain, earrings, or bracelet.

Listen to Your Heartstrings Ear Climber

MATERIALS

3D pen

Filament in a color of your choice

Scissors

Pliers

French wire earring hooks

They say that you should listen to your heart, so this project brings the heart right up to your ear. If hearts aren't your thing, use the lightning template on p. 48 for another fun take on the concept, or create your own shape and wear it with pride.

TO MAKE THE EAR CLIMBER

1. Copy the template on p. 48 onto plain paper. Using the largest nozzle on your 3D pen, begin extruding the filament—this will give you a running start—and then press the tip of the 3D pen to your template. Trace the template using a slow and steady pace. Keep the nozzle at a slight angle and your hand elevated above the work surface. Continue ¼ in. beyond the end of the template; you will trim this later.

2. Inspect your work and touch up any areas as necessary. Once the filament is completely cooled, gently peel the design from the template and trim away the excess filament with the scissors.

3. To make the ear climber as shown, follow the process for preparing the French wire earring hooks on p. 5.

4. Open the loop with pliers and slide it onto the end of the heart. The hook should be perpendicular to the ear climber.

5. Gently squeeze the loop closed about ¼ in. from the end of the ear climber so that it securely holds the ear climber in place. Be careful to not squeeze it too tightly, as this may bend or damage the design. To wear, insert the hook into your ear as with any other earring and then gently angle the design so that it rests along the edge of your earlobe. You can use a rubber stopper, if desired, to hold the earring in place.

variation

The When Lightning Strikes Ear Climber template on p. 48 can be used to create adorable lightning bolt-shaped ear climbers instead of these hearts. The steps are the same; simply follow them as you would to create the heart ear climber.

Interlocking Triangles Necklace

MATERIALS

3D pen

Flexible filament in a color of your choice

Scissors

Jewelry cement or strong glue

Pliers

Two 4mm jump rings

4 fold-over cord ends

18–20 in. of bulk cord or chain

Abstract designs and geometric shapes are the perfect choice for a stand-out statement piece that you can create in a snap. Paint this design in a shimmery metallic for an elegant look, or use a bright filament for something that really pops.

TO MAKE THE NECKLACE

1. Copy the template on p. 48 onto plain paper. Using the largest nozzle on your 3D pen, begin extruding the filament—this will give you a running start—and then press the tip of the 3D pen to your template. Trace the template using a slow and steady pace. Keep the nozzle at a slight angle and your hand elevated above the work surface. Overlap the 2 legs of the triangle at the center and continue on to finish the design. Continue a bit beyond the end of the template; you will trim this later.

2. Inspect your work and touch up any areas as necessary. Once the filament is completely cooled, gently peel the design from the template and trim away the excess filament with the scissors.

3. Repeat steps 1 and 2 so that you have 2 triangles for the necklace. To interlink the triangles, gently pry apart the crossed legs of one of the triangles and interlink it with the other. Using flexible filament or ABS makes this step easier to accomplish without breaking the triangle. Use jewelry cement or strong glue to close the triangle that you opened.

4. Use pliers to attach a fold-over cord end to one leg of each triangle, and also to each end of your cord. Thread a jump ring into the eye of the fold-over cord end on each end of the cord, and then link each jump ring to one of the triangles. See p. 4 for more detail on working with bulk chain or cord.

variation

It's easy to make this design into a bracelet instead of a necklace. Just use a photocopier to reduce the template by 50 percent and follow the steps above, using a cord cut to fit your wrist comfortably.

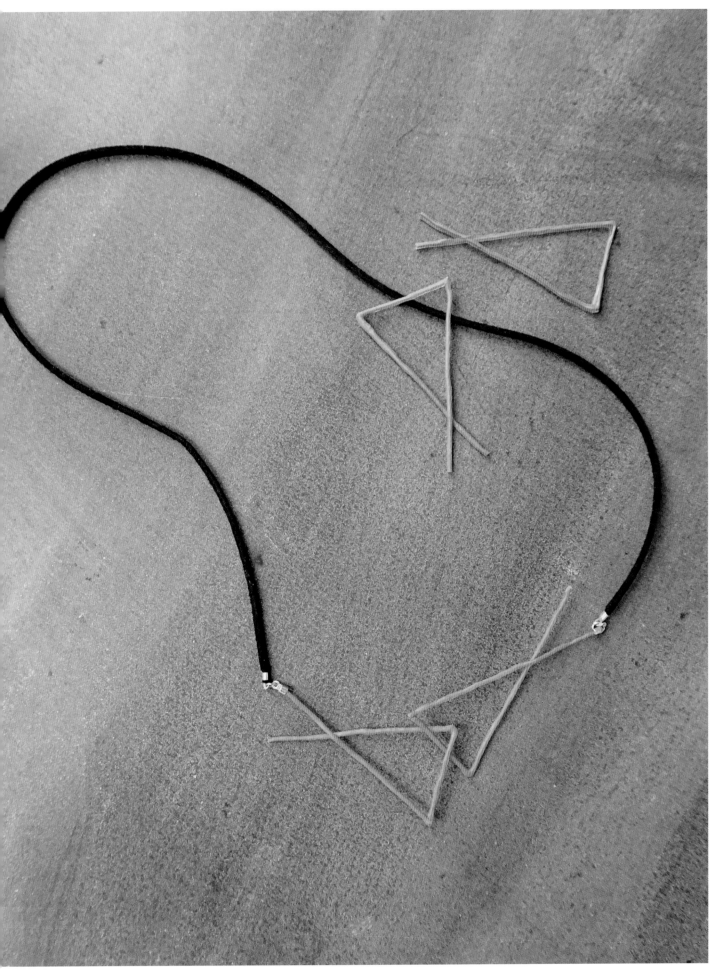

Heartbeat Bracelet

Go ahead—wear your heart on your sleeve. This bracelet is a great way to share how you feel. Space the beats out and wear it when you feel calm, or bring them closer together for a design that is bursting with energy and excitement. Swap them out or stack them as the mood strikes you.

MATERIALS
3D pen
Flexible filament in a color of your choice
Scissors
Binder clips
Cord (optional)
4 fold-over cord ends (optional)
Two 4mm jump rings (optional)

TO MAKE THE BRACELET

1. Copy the bracelet template on p. 48 onto plain paper. Measure your wrist and mark the template to this length. If you need to make the template longer, just repeat the design until you reach the desired length. Using the largest nozzle on your 3D pen, begin extruding the filament—this will give you a running start—and then press the tip of the 3D pen to your template. Trace the template using a slow and steady pace. Keep the nozzle at a slight angle and your hand elevated above the work surface. Continue a bit beyond the end of the template; you will trim this later.

2. Inspect your work and touch up any areas as necessary. Once the filament is completely cooled, gently peel the design from the template and trim away the excess filament with the scissors.

3. To make the bracelet as shown without any hardware, hold the ends of the bracelet together and clip them to your work surface with a binder clip; there should be ¼ in. of open space between the ends of the bracelet. Then, attach the ends of the bracelet by laying down a short segment of filament. Be sure that the bracelet is turned inside out before you connect the ends, or else the connection will show and it will not wear correctly. Let the filament cool.

4. To make the bracelet as shown with hardware, use a shorter segment of the heartbeat template; 3 to 4 beats is plenty. Then, cut a length of cord that, when added to the section of heartbeat design, will match your wrist measurement. Attach a fold-over cord end to each end of the cord, and also to each end of the heartbeat design and crimp in place. Use pliers to open a 4mm jump ring to attach one end of the heartbeat design to one end of the cord, and repeat on the other side. The bracelet should comfortably slide onto your wrist.

variation

Use the earring templates on p. 48 to create a set of matching earrings. Simply trace the template and prepare the ends of your French wire earring hooks as you would for the ear climbers (see p. 36). Slide the end of the design into the loop of the earring hook and carefully close the loop.

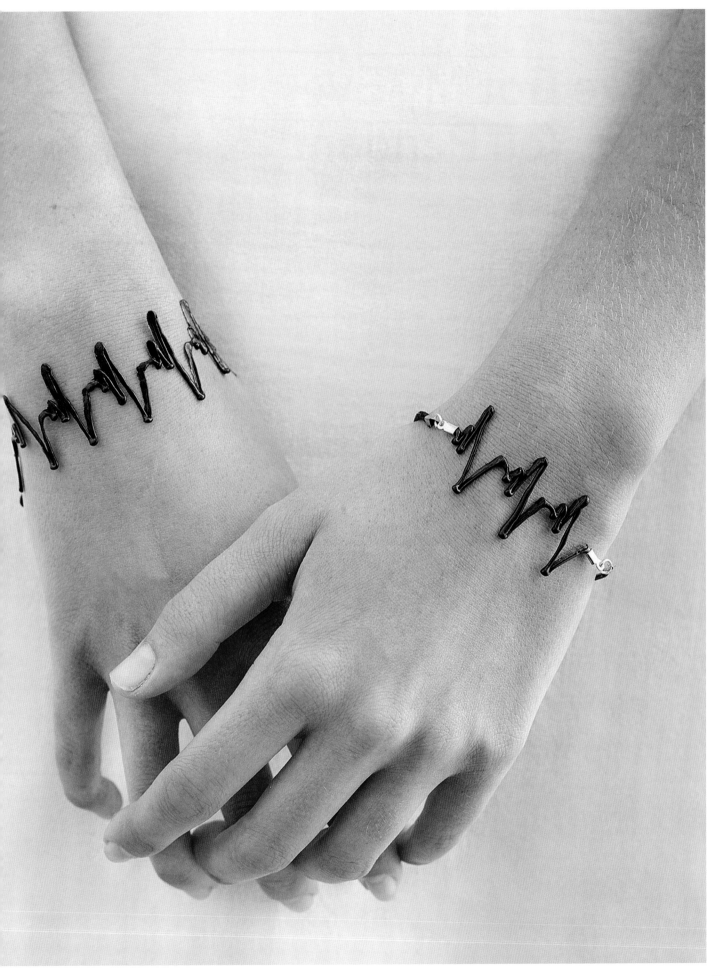

I've Got My Eye On You Pendant

This eye pendant is a bold piece and a modern interpretation on the "evil eye" design that appears in cultures around the world. Even if your eyes are closed, with this pendant you'll always have an extra to watch your back.

MATERIALS

3D pen

Filament in a color of your choice

Metallic paint in a color of your choice (optional)

Scissors

4mm jump ring

Necklace chain or cord with clasp

Pliers

TO MAKE THE PENDANT

1. Copy the template on p. 48 onto plain paper. Using the largest nozzle on your 3D pen, begin extruding the filament—this will give you a running start—and then press the tip of the 3D pen to your template. Trace the inner circle of the template using a slow and steady pace. Keep the nozzle at a slight angle and your hand elevated above the work surface. When you return to the starting point, continue extruding past the edge of the circle; this will give you a stronger closure to the circle and you will trim it later.

2. Once the filament has cooled, trim away the excess filament without removing the circle from the paper. Trace the outer portion of the template, making sure it rests snugly against the inner circle at the top and bottom (as shown).

3. Inspect your work and touch up any areas as necessary. Once the filament is completely cooled, gently peel the design from the template and trim away the excess filament with the scissors. If you would like to paint the pendant in a metallic finish, now is the time to do so; I used a silver paint.

4. To make the pendant as shown, open a jump ring with pliers, place it around the top of the eye (as shown), and close. Thread a chain or cord through the jump ring to complete.

---

variation

To make a chic bracelet instead of a pendant, simply attach a jump ring to either end of the eye and finish with a length of chain or cord that comfortably fits around your wrist. To make a ring, use a photocopier to reduce the template by 50 percent and attach the eye to a ring blank using jewelry cement.

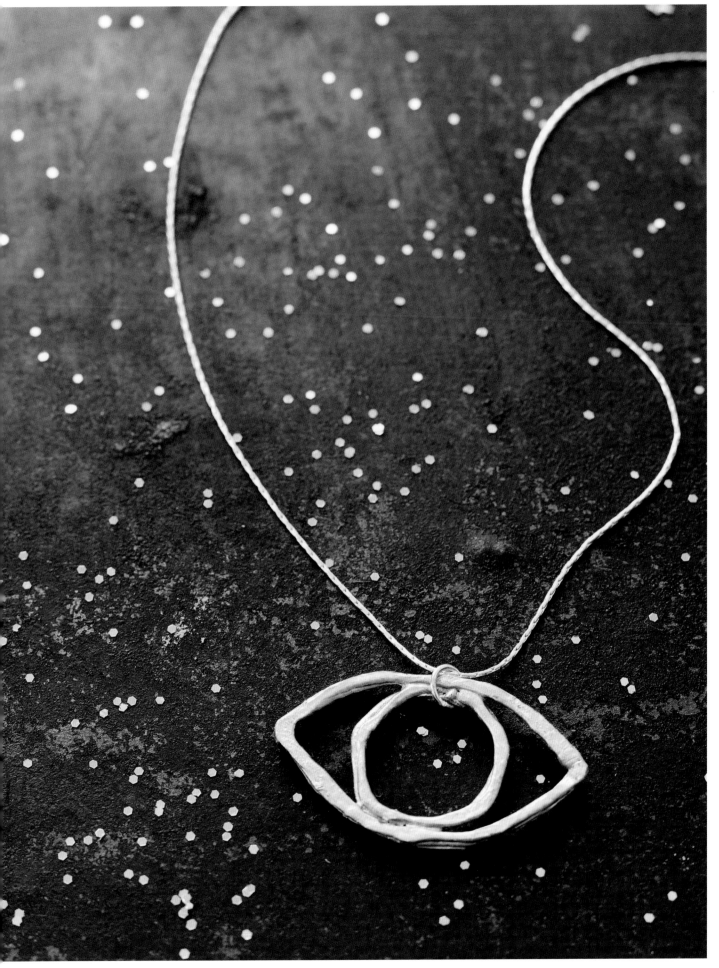

Templates

CRESCENT MOON PENDANT P. 6

MODERN CORAL LARIAT P. 10

THE CAT'S MEOW RING P. 8

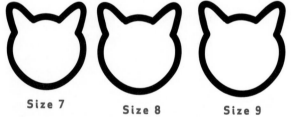

Size 3   Size 4   Size 5   Size 6

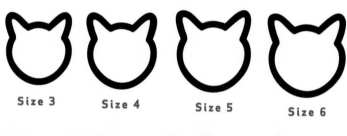

Size 7   Size 8   Size 9

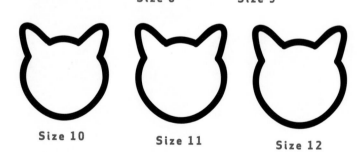

Size 10   Size 11   Size 12

INFINITELY CIRCULAR EARRINGS P. 12

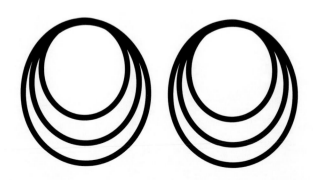

OMG

XOXO

YAY

LOL

WHAT'S YOUR SIGN PENDANT P. 16

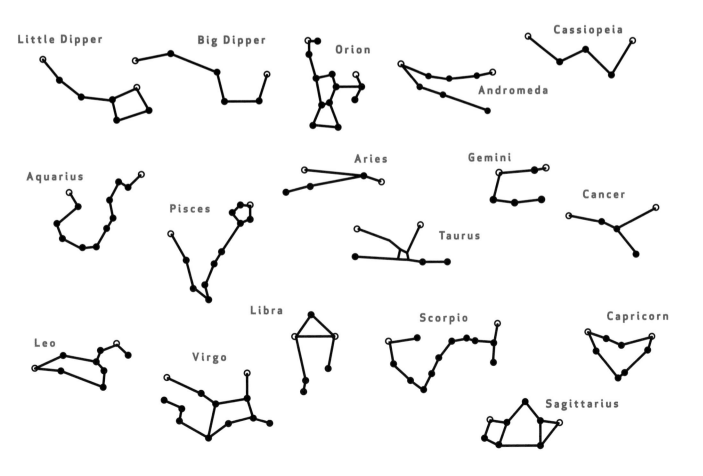

Little Dipper

Big Dipper

Orion

Cassiopeia

Andromeda

Aquarius

Aries

Gemini

Pisces

Cancer

Taurus

Libra

Scorpio

Capricorn

Leo

Virgo

Sagittarius

WHEN IN ROME BANGLE P. 20

RUFFLED FEATHERS NECKLACE P. 22

ALL COILED UP RING P. 24

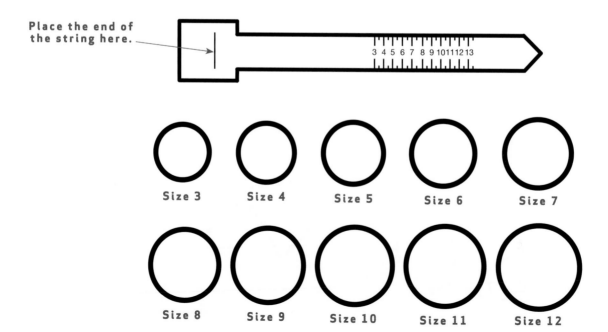

Place the end of
the string here.

3 4 5 6 7 8 9 10 11 12 13

Size 3   Size 4   Size 5   Size 6   Size 7

Size 8   Size 9   Size 10   Size 11   Size 12

GEO-TOTEM NECKLACE P. 26

IT'S MOLECULAR PENDANT P. 28

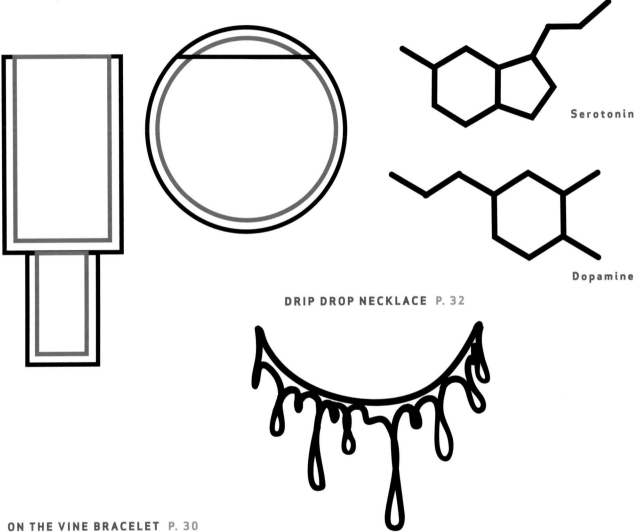

Serotonin

Dopamine

DRIP DROP NECKLACE P. 32

ON THE VINE BRACELET P. 30

Bracelet

Necklace

DISCARD

SAY IT OUT LOUD NECKLACE P. 34

LISTEN TO YOUR HEARTSTRINGS EAR CLIMBER P. 36

WHEN LIGHTNING STRIKES EAR CLIMBER P. 36

totally

girly

love

sight

hope

dreamers

Maybe

INTERLOCKING TRIANGLES NECKLACE P. 38

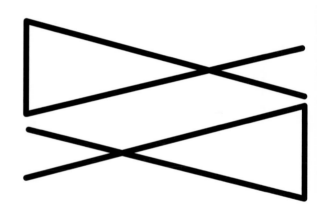

I'VE GOT MY EYE ON YOU PENDANT P. 42

HEARTBEAT BRACELET P. 40

Bracelet

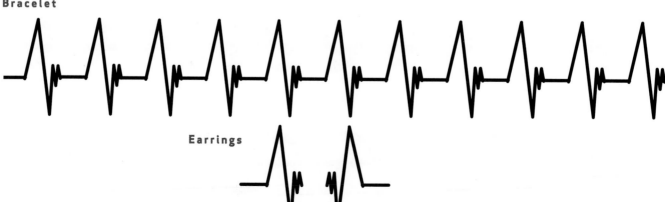

Earrings